Publisher
Interstellar

Editors
First editor: Adaora Uche **(marthashands.hm@gmail.com)**
Second editor: Lola Denton **(codeladylee@gmail.com)**

Cover Design
Akinlabi Akinbulumo **(www.phishaman.com)**

Formatting
Akanimo Inyang **(www.excite.ng)**

Dedication

To my 17-year-old-former self, and to everyone who wishes to find an alternative source of livelihood in this adulting stage called life. You will do great things, and deserve it.

To my wife Ivie and children, Adrian and Andre. You are the source of my strength and my constant north-star, my "localhost". Ivie, thank you for constantly requiring the best version of myself in anything I do.

To my parents. Dad, Mum you have loved me unconditionally and your examples have taught me to work hard for the things that I aspire to achieve.

And finally to God Almighty, my maker. The source of my inspiration, wisdom, knowledge and talent. You have given me seeds, which I will continue to multiply and share as long as you will it.

Acknowledgment

I want to start by thanking my wife and queen, Ivie (dtooth-doc). From ideation, to reading early drafts, bouncing topic ideas and finally committing to doing this. As always, your support, friendship and quality control measures kept me grounded through this process. Thank you so much darling.

To Bankole "Lordbanks" Olufemi, you actually planted the initial seed all those years ago during one of our many conversations, which helped put me on this path to writing this book. I didn't think I had anything to give, but all you said was "It's not that hard: Just write out your chapter titles", and here we are :). Thank you.

To Bankole Williams, your inspiration, counsel and direction throughout this process has been impactful and very helpful. Through our long sessions over the phone to even revising to a more befitting book title. Brother, I greatly appreciate all of this, thank you.

Thanks to everyone on the Book team who helped me so much. Special thanks to Ada Uche who was significant to my editing process, Lola Denton for your second phase editing and final form, Akanimo Inyang for the amazing book outline/layout and Akinlabi Akinbulumo, for the kick ass cover design and brand outlook for this.

You all are amazing, thank you.

Chinaza Obiekwe, thank you for the final stretch!

To all the individuals I have had the opportunity to lead, mentor or impact in one way or the other, I'd like to say a big thank you for being the inspiration and foundation for writing this book. My hope is that this book will help you or someone in your ecosystem out there.

Foreword

by Victor Asemota

I was already a young adult before I started using a computer for the first time. It changed not just my future but also that of the many colleagues who later joined me to launch several technology companies.

Interestingly, my first technology job was in a training role. I was responsible for training people who had no tech backgrounds on how to use computers to run organizations like banks and even oil companies. There are many possible career opportunities in technology, and you can start from anywhere and become anything you want to be.

I didn't study information technology formally in school until my second post-graduate course in the UK. The first thing I also learned there in my very first course module was that information technology exists to serve people and not otherwise. "There is nothing like a technology project; there are only business change projects that apply technology."

Many people are not fortunate to have started early the way Chuka started by learning by himself after his mother gave him her computer to play games with after teaching him the basics. His understanding from an early age nevertheless has helped him get to the top in his field as one of the top experts. I am delighted that he has decided to share his knowledge with all those interested in joining us as technology practitioners.

I met Chuka when he was a co-founder of the tech startup Delivery Science. They were part of the first African cohort at the Google Launchpad Startup Accelerator in San Francisco. I was immediately impressed by the time he always took to explain and break things down when asked questions. He became a Google Developer expert and one of the most respected technical mentors for the African google for Startups Accelerator. There is no person better qualified to write a book like this.

Africa faces a massive digital skills gap. According to the World Economic Forum, "Demand for digital skills training in Africa will surge in the coming decade as jobs that before did not need digital skills will begin to do so." This demand means that technology will become ubiquitous, and we have to prepare for it.

This book is not just a career guide; it is probably going to be the key that opens the minds of many more young Africans to the opportunities in the technology industry and the application of technology into everyday life and business. I hope it will change lives the way technology transformed my own life and others by extension.

Contents

SCAN ME

Introduction

The decision of what career to pursue is one that is oftentimes made very early in life. One question children get to be plagued with is that of what they hope to become in future. More often than not, their bubbly minds always have an answer: doctor, musician, lawyer, etc.

I have also interestingly discovered that over the span of their childhood, several career choices appeal to many of them: first from wanting to be in the Navy, to wanting to become a lawyer, and then a rock star etc. They just want to be it all. Life is vastly full of potential and opportunities that cannot be exhausted. Many times, these uncertainties in a career choice that children exhibit tend to make parents seek to help them make a decisive choice - after all, the furtherance of their academics depends on a specific path.

So, the child chooses a specific path and then begins to pursue it. Interestingly, life begins to call out again with that gnawing, familiar and unsettling desire to have and be much more, riddled with so much uncertainty.

The only problem now is that as adults we ought to know what we want, stick to it, build it, grow it and then reign comfortably. At least that is how our minds were conditioned. But this is hardly so. As we mature and grow older, we lose our childhood zest for wanting to be everything we could ever be, replacing it with the desire to be settled and established. Changes to this settlement we have attained are often unwelcome, and most often, totally resisted by most people. Yet, when change comes knocking, it must be answered either willingly or unwillingly.

A career transition is one of the most uncomfortable transitions one could experience in life.

Understanding Career Transitions

Transitions are a part of life. They happen in all aspects of life, be it physically transitioning from a child to an adult, from single to married and then possibly a parent, or losing the presence of some of your loved ones. Transitions also happen in career too, no doubt. A career transition is one of the most uncomfortable transitions one could experience in life.

Generally, people expect an upward linear progression (simultaneously or exponentially) in their careers. Unfortunately, this ends up being untrue for many people, save for some.

Moving along the career path, two major events could happen to cause a shift which may lead to career transitions:

1. **Loss**

   For many professionals, experiencing loss and its aftermath in their career provokes a re-evaluation. Losses could come in the form of personal or job loss, denial of promotion, opportunities or even due recognition, etc. The result of this re-evaluation could unbox a myriad of revelations, one of which could spark a desire to switch lanes career-wise. For instance, the loss of a job could help the affected person realize that they have always possessed an entrepreneurial spirit or perhaps a flair for IT which they had previously taken for granted and now must resort to, just to make ends meet. Loss of opportunities or due recognition could lead to a downward spiral of negative emotions for the affected individual, which could provoke the need for a change in the way work is done, or a complete loss of interest in the current job. Thus, such a person may prefer to seek for other career opportunities or to embark on a completely different career journey.

2. **Appetite Changes**

   As it has been earlier established, more often than not career is not a linear journey. This is because aside from the disappointments or delays that could happen, people's appetites change on the job and over time. There are extreme instances of people who studied a particular profession in school, commenced a career in a different field and along the line discovered they would rather do something entirely different. This is quite similar to the "divergent" career interests of a child who desires to be everything. As we progress in life, we tend to change in tastes, priorities and desires. A strong thirst for more is one common trait among the human race: a thirst for more recognition, fulfillment, contribution, reward, remuneration, etc. This desire often provokes changes to be made - one of which could be a career shift.

Making a career transition is never easy and requires a lot of planning, preparation and positioning. In some fields, it is very difficult to transition because of the high level of specialization needed. Transitioning to Tech is wrongly perceived to be one of such because of its highly technical nature. This book is designed to demystify this fallacy and help you transition and settle into a career in Tech, either as a technical or non-technical professional in the Tech ecosystem.

SCAN ME

Chapter 1

Technology is the New Oil.

Technology has always been the driver of all industries but has only always been given a place in the booth because of its cumbersome and complex nature. It never was considered as a revenue-generating industry by itself, but as one that needed to be invested in so that other industries could generate more revenue through its dividends.

This classification as an indirect revenue generating industry is what caused the African continent, the original progenitors of technology, to ignore investments in Technology as a sector. Africa focused more on the agriculture, commodities and natural resources sectors, choosing instead to import technology to drive its economy. Today however, with the advent of the Information Age, technology as a sector has taken its rightful place among other sectors of the economy all around the globe.

Let's back up a bit.

In the 1800s, the oil and gas industry was born in Pennsylvania in the United States of America, thus changing the economic landscape of that nation for the better. A lot of countries with large oil deposits benefited from trading or refining crude oil and natural gas as a commodity, Nigeria being one of them. Thus, for a long time, the oil and gas sector remained unrivaled as one of the world's biggest industries, with the oil drilling sector alone contributing 3.8% to global revenue. "Black Gold", as oil is also called, became pivotal to the functionality of many sectors of economy such as power, aviation, transport and logistics, manufacturing, space engineering, etc, as its products and by-products found relevance in almost every industry.

This beau ideal was poised to be shaken, unfortunately. Its downturn started as a result of the consequent incessant pollution and toxicity to the environment (land, sea and air), which escalated from being just a global concern to a global environmental hazard that needed urgent mitigation. Additional problems such as crude oil being exhaustible, hazardous, conflict-provoking between countries and non-renewable led to the need for more sustainable sources of energy (such as renewable energy).

Technology stepped right in again to proffer the much needed solutions. The previously quiescent Technology became more assertive and ready for global dominance, no longer to be relegated to the background while other sectors took the shine.

You see, the "Gentle Bull" (Technology) was finally awake, and though he had penetrated every sector as always, this time he walked in boldly, right through the front door with his irresistible solutions. He moved from an obscure place in the trunk to the driver's seat. The Gentle Bull had always been the "oil behind the oil", as there was no oil and gas industry (or indeed any other industry) without Technology. He had always been at the heart of all industries, but now he chose not to be just a part of every industry, but to be the core of every industry. Today, every industry is a "Technology industry", focused on specific services such as aviation, health, agriculture, finance, etc.

The invention of and huge advancements made in computer technology have in no small way accelerated the rate of impact Technology has had globally especially in this 21st century, setting the stage for the Fourth Industrial Revolution. However, with the recent Corona-virus pandemic that has plagued the world since its discovery in Wuhan, China in 2019, the global economic stage has been doubtlessly seriously impacted. The corresponding efforts to curb the spread of the Corona-virus disease such as social distancing, lock-down restrictions, and bans on international travel have led to the rise and fall of several industries.

This has in turn caused the digital economy birthed about three decades ago, which was steadily on the rise, to suddenly experience an exponential leap in its growth. Today, the previously regarded indirect source of revenue is not just a sector but an economy of its own. The viability of any industry in these times is largely dependent on its synchrony with this digital economy.

Most certainly, with remote work being the new face of work, technology has never been more powerful or indispensable. The need for technology solutions has skyrocketed, creating a larger market in the Tech space, numerous opportunities for work in the ecosystem and an unprecedented boom in the digital economy. This, in effect, means that Tech is the new oil - not only for economies of nations, but also in the career space. With so many exciting perks such as an ever expanding industry offering numerous employment opportunities, above average remuneration as well as its significant contribution to humanity, it is no wonder Tech is the new oil. In Nigeria, for instance, oil and gas sector jobs are so desirable because of the great remuneration associated with them, especially in the top global exploration companies. The next industry that offers such great benefits is the Tech space, Information Technology (IT) to be precise.

Many people have an eye for the industry but are often discouraged by the high level of technicality and expertise it seemingly requires. Also, many have a limited understanding of the Tech industry. They view it with the microscopic lens of solely software and services however, the industry is a very vast one with numerous fields such as Telecommunications/Internet Service Provisioning, Cybersecurity, Technology Manufacturing, Infrastructure as a Service (IaaS), Platform as a Service (PaaS), Artificial Intelligence, Business Data Engineering etc.

Good news! Seeing as it is such a huge industry, IT requires different kinds of professionals. Yes, you heard me: **you do not have to be technically inclined to work in Tech.** Stick with me as I show you how to transition to a career in Tech despite knowledge limitations.

Did you know?

Silicon Valley residents out-produce almost every nation on the planet. According to the Federal Bureau of Economic Analysis, California's Tech Belt yields an output pegged at $275bn, with an annual GDP of $128,308 per capita, which is higher than Finland's GDP.

SCAN ME

Chapter 2

Technical or Non-Technical?

There is the story of a teenage girl who took up computer programming as a hobby due to her obsession with an online game she loved to play. However, she dumped this hobby to pursue a degree in Psychology, after which she went on to bag a master's degree in Law. It was during that pursuit she realized that her favorite pastime (computer programming) was indeed what she wanted to pursue as a career. Thus, she made the switch, teaching and training herself via online courses and blogs. Today, Ire Aderinokun is the first female Google Developer Expert in Nigeria. She has gone on to empower more young ladies to take up more seats in the Tech space by sponsoring them for IT training in software development.

Contrary to what many believe, Tech is one of your best second-chance options at a career. Whether you have a degree

“

Life always offers you a second chance. It’s called tomorrow.

Dylan Thomas

in Political Science or Insurance, Tech could be for you. You can get into Tech and have a fruitful career as a technical professional without having a Tech background academically. You could pick up technical skills and knowledge - not very easily though, but very possibly.

With Tech, you are not restricted by the age bias many professions set as restrictions. You can enter the industry at any age. It is a function of what unique skill sets you can bring to the table. I know you could argue that this is not really a big deal because of anti-ageism laws, but we know that these laws are hardly respected.

A good example is a typical Nigerian graduate who has had to write JAMB (quite similar to SAT exams) for several years before securing admission to study in a higher institution at an average age of 19. Then comes an ASUU (academic staff union) strike that causes further delays for at least a year, or sometimes two, giving an average of 24 years at graduation. After this is a compulsory one-year conscription with the National Youth Service Corps (NYSC). It takes an average of four to six months before being called up for the service year, a prerequisite for being eligible to be in the labor market, which primarily consists of the Banking sector - the most employable sector in Nigeria before IT - and then others. The sad reality is that these banks usually place a maximum age limit of 24 years for entry-level jobs, whilst the average graduate is already 26 years old or more and getting even older while waiting to secure a job.

The difference with IT is that nobody really asks what your

age is before employing you. They are more interested in what skills you have to offer. This is not only true for IT as a sector but for IT roles across other sectors. It is always about solutions and competencies first, not age. The younger you are though, the easier it could be for you to catch up with this fast-paced industry which requires a lot of brain power. Nonetheless, opportunities abound for you, age notwithstanding. I once read a blog post by a woman who, after 20 years working in logistics and then linguistics, realized at the age of 43 that her fulfillment was in Tech. She then took some software coding classes, and after several months of freelancing and networking (I will share more on this in the later part of this book), she got her first job in Tech. The beautiful thing about it was that Tech offered her the opportunity to combine her 20 years work experience with her new job as a Customer Support Manager. Three years later, she became the Operations Manager of the same company. Recall that she had no Tech background, just an interest - and was over 40 years of age.

According to the U.S. Bureau of Labor Statistics, "Employment in computer and information technology occupations is projected to grow by 11 percent from 2019 to 2029, much faster than the average for all occupations. These occupations are projected to add about 531,200 new jobs. Demand for these workers will stem from greater emphasis on cloud computing, the collection and storage of big data, and information security."

Secondly, in the Tech space, you do not have to work a technical job to have a lucrative career. According to Glassdoor

Economic Research, 43% of available jobs in the Tech industry are non-tech jobs. In a hospital for instance, despite the number of medical professionals working there, non-medical staff are also needed for other administrative and business related activities. Similarly, there are technical and non-technical staff in every Tech company who handle admin, finance, logistics, sales, product development etc. This will be discussed in detail further in this chapter.

Thirdly, the remunerations for working in Tech, especially as a technical professional, are quite enviable. In an article by Fox Business, it was stated that Tech jobs are among some of the highest paying jobs in the world.

Another strong point is that Tech offers you flexibility with work because of its many remote work options. Living in a hectic city like Lagos where, on the average, five to six hours could be spent commuting in traffic daily, many workers find it difficult to maintain work-life balance. However, working a Tech job could to a large extent help ease that as virtual work options have been available in many Tech companies in Nigeria long before the global COVID-19 pandemic and its consequent lock-downs and social distancing restrictions.

Lastly, Tech offers you one of the shortest transition times possible, as compared to many other professions. Depending on the area of interest you indicate, you could transit to Tech in as little as three to six months, although this would depend largely on your level of consistency and dedication to bridging your knowledge gaps. It is best to take your time when seeking to transition to a new career. It is a major life decision and must be met with the highest level of preparation possible.

To transition to Tech, you must first decide what you really are after - lest you get really frustrated in this ever dynamic industry which demands extra-long hours, continuous learning, specialization and certifications, etc. As nice and attractive as it may seem, many professionals in Tech also hate their jobs and are seeking to transit to other sectors. You need to decide if you are interested in Tech just as an industry with your current non-technical skill set, or interested in Tech as a technical professional. I must also mention that you should also have an interest in the Tech industry even if you decide to pursue a non-technical career, because you are likely to encounter some technicalities on the job that will require some level of foundational knowledge as you go along the way.

But I Hate Coding!

Good news, software coding (programming) is not the only career path in Tech. As stated earlier, there are non-technical roles in Tech, as well as other technical career paths such as data, software, infrastructure, networks, etc. The Tech ecosystem comprises several niches that one could comfortably identify and settle into. The Tech space is so vast that your interests are likely to find a place for specialization with corresponding opportunities. Let's move on to discuss some of the most lucrative roles in Tech (IT).

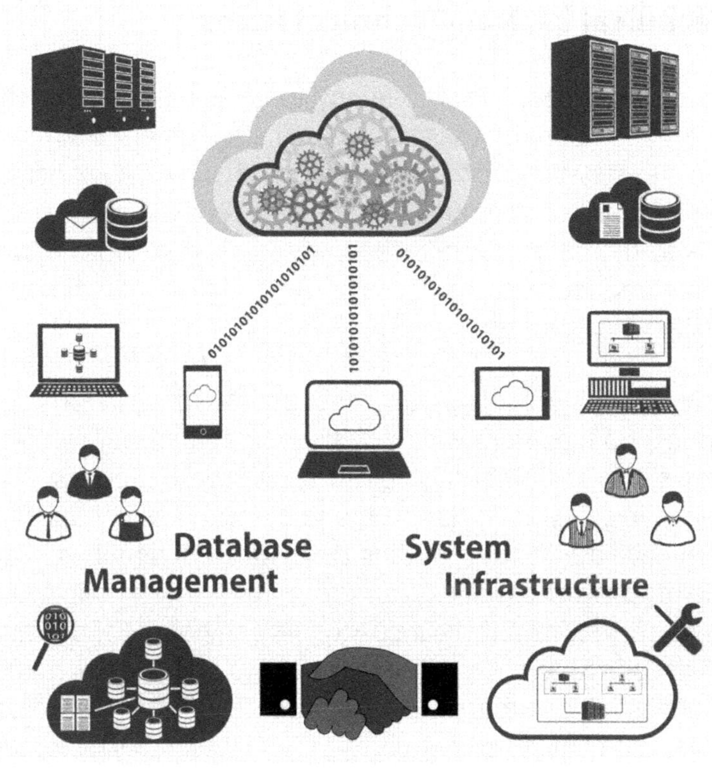

Technology hub showing different nodes & components.

Technical vs. Non-Technical Roles

As earlier stated, every company today is basically a technology company offering services to its customers. However, there exists a core Tech ecosystem which has many spheres. For the sake of optimum learning and benefit, I will restrict this to the Information Technology (IT) sphere which I will sometimes refer to as "Tech" interchangeably within this book.

IT is a stem (hub) with a number of boughs (think in terms of nodes) such as Hardware, Software, Communications, Networks, Cloud Computing, Security, Database, Servers, Applications, Customer Support, and others. These boughs in turn have their corresponding twigs (think of terminus). For the purpose of specificity, and to maximize your gain from this book, I will limit the focus to just one bough – Software Engineering. However, I have tried above to list out as many "boughs" as possible so you could conduct further research if you are convinced that you'd rather transition to another field of IT.

Notwithstanding, I can assure you that this book will guide you and ease your transition into any other field of IT you may choose. I have chosen Software, not only because it is my area of mastery, but also because it is likely the most flexible, receptive and fastest field for a career transition to Tech either as a technical or non-technical professional.

Technical (Tech) jobs are jobs requiring technical skills, knowledge, experience and education to design solutions to business problems. In order to get a Tech job, you need a deeper-than-average knowledge.

Non-technical jobs, like the name clearly implies, are jobs that do not necessarily require a deep knowledge of the technology of your company's business or its relevant skills. However, they are necessary for the success of the organization's operation and strategic objectives.

While technical roles ensure that solutions are designed and services are available for deployment, non-technical roles ensure that these solutions and services reach target customers, commerce is established, customer support, satisfaction and retention are achieved, all the while growing the organization's image and brand.

A good illustration to represent this is an auto salesman and an auto mechanic in a car shop. While the auto mechanic (operating in a technical role) understands the intricacies and operability of the car's engine, he may not have the skills and persona required to sell the car. Thus, the auto salesman (operating in a non-technical role) is the one who can sell the car to interested buyers. Also, the auto mechanic (who is too busy solving faulty issues with a car) would be over-burdened if he had to handle finances, render customer support, market the car shop, make sales, and be responsible for growing the brand of the company whilst attending to legal matters that may arise.

This is exactly the symbiosis that exists in the Tech space between tech and non-tech departments, and why there are lucrative opportunities in sectors worth billions of dollars for even non-tech professionals. All hands are constantly on deck to build success. Transitioning to Tech requires that you carefully make a choice by considering your interests, experience, skills and knowledge.

Technical Roles in Tech

For those interested in Technical roles, there are a world of opportunities, but I will restrict my scope to the software industry, where you could pursue a career as any of the following:

1. **Front-end Developer:**

   As a front-end developer, you will be working on implementing the end-user facing software for any application or website. This could involve the popular User Interface/User Experience (UI/UX) design as well as web design, but it more so requires some serious coding and debugging skills to make the website or application responsive and functional. However, more serious work is needed if the website or application is meant to interact with a back-end database or server. Skills in programming languages like CSS, HTML, jQuery and JavaScript are necessary for this role. You will require knowledge of Internet browser trouble- shooting, debugging practices and techniques, search engine optimization (SEO) principles, and graphics design software such as Adobe Suite and Photoshop.

2. Back-end Developer:

As a user reading a blog for instance, all you see and can relate with was designed by a front-end developer. However, there are background processes and protocols that ensure the functionality of this web-page. The information you fill in an online form for instance, has to be stored in an unseen back-end database. What ensures that your filled in data goes from the web-page you are interacting with to the right back-end server is the responsibility of the back-end developer. They ensure that the interface built by the front-end developer is fully functional, relational and responsive with its linked database, servers and browsers.

In more technical words, back-end developers are usually responsible for writing the web services and Application Programming Interfaces (APIs), which are data endpoints used by front-end developers and mobile application developers for all user-interfacing elements. As a back-end developer you will be responsible for server-side web application logic as well as the integration of the front-end part.

This will require skills in programming languages such as C/C++, Java, PHP, Ruby on Rails, Python and .NET as well as front-end programming languages such as HTML, CSS and JavaScript. You also need to acquaint yourself with program control management systems like Git, GitHub and Docker.

3. **Mobile Application Developer:**

   We use so many applications (or apps) today, especially on our mobile devices. These apps are built by mobile application developers for platforms such as Android, iOS and Windows. These developers create, test and program apps for computers, phones and tablets.

   As a mobile developer, you will likely work on everything the front-end and back-end developer does, but in the target programming language for the platform in question. For instance, in Android development you will need Java knowledge; for iOS development, you will be required to know Swift, and for Windows you will be required to know C#. In addition, as a mobile application developer, you require strong analytical and problem-solving capabilities - just like in the case of a back-end developer. You will interact a lot with applications and languages such as Java, Java EE, Java ME, JavaScript, JSON, Objective-C, .NET, Swift and HTML, depending on the target platform.

4. **DevOps Engineer:**

   DevOps in software engineering bridges the gap between the developers and other IT staff such that production (live) deployment and usage of a product or software application moves seamlessly across board: from the developers to the IT staff that will support the application. With DevOps, organizations can release small features very quickly and incorporate the feedback which they receive equally quickly. DevOps works to ensure fewer software failures, shortened lead time between fixes and better management of the software development life-cycle. Your role as a DevOps engineer will be similar to that of a project manager. Some tools you will work with are Kubernetes, Docker, Git and GitHub, Jenkins, Selenium, cloud resources (such as Google, MS Azure and AWS) as well as other programming languages stated above.

5. **Full-stack Developer:**

    Full-stack developers, as the name implies, are computer programmers who are proficient in both front-end and back-end coding (both stacks for front-end development and back-end development). Their primary responsibilities include designing user interactions on websites, developing servers and databases for website functionality and coding for mobile platforms. Generally, a full stack developer is required to possess skills that both the front-end, back-end, and even mobile developers possess. Hence, it is a highly sought-after skill in the software development world.

6. **Software Architect:**

    As a software architect you are likely to lead a software development project team and will determine which processes and technologies a development team should use. You will interact with businesses and clients to design and execute solutions for them, working with a team of software engineers whilst reporting to senior management. Software architects design and develop software systems and applications.

You may be involved in creating customer specific software solutions or general consumer products such as games or desktop applications. This is a high-level decision-making role which requires deciding everything from design choices to technical standards such as platforms and coding standards. Relevant competencies needed to transition to a software architect role are research, Agile project management, team management, client communication, quality assurance and definitely, knowledge of coding.

7. **Big Data/Data Engineer:**

The world has an insatiable appetite for data. Data is constantly being generated by all the other practices discussed above. Data engineers are good at programming, with great emphasis on specialization in distributed systems and big data. Big data engineers are professionals who process massive data sets to provide their organization with detailed analyses that can further be used to make future decisions or to avoid mistakes in the future. As the world evolves, "Big Data" continues to grow and evolve. New functions and specialties arise, and specialized positions emerge.

Data engineers pursue insights, truth, reliability and delivery through accurate data analysis and representation, for example, a machine learning engineer is a kind of data engineer. A data engineer should have about the same skill set as a DevOps engineer which will also include SQL, data acquisition, data gathering, data cleaning, and ingesting data, experience with schema design using semi-structured and structured data structures, implementing data pipelines via the ETL (Extract, Transform, and Load) model amongst a host of others.

8. **Machine Learning Engineer:**

   This role spans across two domains, namely data science and machine learning hence it can be said to bridge the gap between both fields. Machine learning engineers come from data engineering backgrounds, but also have enough experience to be proficient in both data engineering and data science.

   The difference is that while data scientists extract meaningful insights from large datasets and communicate the information to end users or consumers in tables or charts, machine learning engineers ensure that the models used by data scientists can ingest vast amounts of real-time data for generating more accurate results to be either used as standalone data or in a software that has been developed by a front-end or back-end engineer. It is important that a machine learning engineer should be able to create simple and possibly complex servers for the sake of quick prototyping.

   For a career in this field, you will need knowledge of computer science and programming, probability and statistics, data modeling and evaluation, machine learning algorithms and libraries, software engineering and system design.

Non-Technical Roles in Tech

In the same vein, if you are interested in Tech as an industry without being a technical professional, there are many opportunities for a career in the industry. We shall explore some of them:

1. **Sales:**

   If you possess great interpersonal skills and are a great salesperson with a great track record of sales, you may want to consider a technical sales role. While this is not a very technical position, you will require good understanding of the technology behind the product you would be selling.

   Most Tech companies prefer to have people with a technical background or experience in this position, but hey - the beautiful thing about being a salesperson is being able to sell water even to a fish. If you can convince your job interviewer that you can sell their product regardless of your background and guarantee them an increase in their sales volume, an increase in their sales volume and market share based on your previous exploits, you just may land the job. You will need to get right into understanding the technology of your new company's business, and quickly too. You will need to liaise with clients, organize sales visits, prepare tenders/

quotations, provide pre-sales and post-sales support, review sales performance, provide product education and advice etc.

2. **Project Manager:**

   One feature that cuts across the Tech industry be it software, telecoms, IT infrastructure etc, is projects. The technology industry thrives by deploying multiple solutions to multiple clients. Companies run numerous projects per time. You could aim for a project management role without needing to even understand a thing about the technical requirements of the business. Depending on your background or interests, be on the lookout for roles such as schedule management, cost management, HR, process writing, quality management, contract management, bid/engagement management, performance management, etc. Having a professional certification (such as PMP) or an educational qualification (such as an MBA) could give you an added advantage in landing this role.

3. **Product Manager:**

   The business of technology is in creating products and offering services as solutions. This necessitates the management of products created. Product management is about overseeing and managing a new product from conception through its development life-cycle, and finally to its commercial launch and sales, all the while ensuring that the product is in alignment with the company's overall strategy and goals. Technology companies require product managers who will define the product vision, strategy and road-map. They work closely with the product developers, engineers, operations, deployment, marketing and sales teams to ensure the product is created according to customer requirements and specifications.

Business developers and analysts would also be needed in product management to ensure that products can generate new business opportunities, minimize operational and running costs in product development as well as ensure that the product can thrive in the market ensuring increase in profit and growth for the company. The product manager role is best for those who enjoy collaborating with others, are good with creative and analytical thinking, and have strong communication skills.

4. Customer Success Manager:

The goal of customer success management, otherwise known as client relationship management (CRM) is to ensure that client expectations are maximized while using the company's products in order to sustain a perpetual and mutually beneficial business relationship. This role requires an overall knowledge of the product and some key competencies such as good communication skills, negotiation skills, relationship management, problem solving skills, etc.

SCAN ME

Chapter 3

Picking the Right Role.

Mayowa and Chris met on their first job and hit it off immediately. The two men often talked about their career aspirations. They both worked as IT network engineers, a very demanding job that required 24x7x365 network surveillance. This also meant they had to work shifts even on weekends and could not have a regular 9-5 work schedule like most people. Though they both really wanted to transition to other roles, it was difficult to plan their lives because of their irregular and tight schedules. They worked six days a week: two days on the morning shift which was expected to end by 2pm (though they never really left work until 4 or 5pm), the next two days on the afternoon shift, and the next two on the night shift. They had two days off afterwards, starting the morning after they got off the second night shift.

Unlike most night shift jobs, working in network surveillance was hectic. One barely had a moment to catch a nap. Due to the intense work pressure, they both soon started losing enthusiasm for their jobs. Their career aspirations were slowly getting buried. Year in and year out, they desired a change, but they had alas fallen into the rut of stagnation common to most career people in that position. Also, they were discouraged - seeing that most of their teammates had been on the same job role for years. Network surveillance was a forgotten department - promotions were scarce if not non-existent, with little or no opportunities for career advancement. They soon joined the rest of the team in complaining and wishing for something better. Work became a compulsory jail sentence they had to serve every day.

One morning, as Chris wearily got out of bed to prepare for his 7am shift, he looked at himself in the mirror for the first time in a long time. *"Oh boy, four years don waka o*[1]. *I still dey this hustle. When would things change for the better?"* He pondered. Strangely, it dawned on him that four years really had gone by, yet he had been on the same spot, wishing for an unknown time T to bring him the opportunities he craved for. That day, he knew that he needed to change things and not just hope that his dreams will come true.

Whilst he worked all through that day, he discussed his thoughts with Mayowa.

"Guy! It's been four years we have been in this role o, it is time for a change", Chris said.

---

1 *Four years have elapsed.*

"Bro, wetin man go do[1] na[2]? One day sha[3], I know God will remember His children", Mayowa replied with his usual resignation.

"Guy, I know say God dey[4] but free that matter first. How can we change things? What do we even want?" Chris asked.

"Na wa o[5], see question o. To change departments of course, or better still get a better job with better pay," Mayowa responded.

Chris considered this statement. "I know, but what department? What job?"

"Guy, any one abeg[6]" Mayowa paused, refocusing on his computer screen, trying to prepare an escalation report for a site outage. "Any other job but here." He said after a few moments.

"Could that be the problem?" Chris pressed. "We want a change, but we do not know exactly what change looks like? What are we really pursuing? Does it have a name? Who does it look like? Where is it stationed? How do we get there?" He went on, speaking more to himself than to Mayowa this time. "That, my friend, is the real problem we have o, not just God and time."

---

1 *What can i do about the situation?*
2 *"Na": used for emphasis, no actual meaning.*
3 *Some day, for certain.*
4 *I am aware that God can help us.*
5 *Are you kidding me?*
6 *My friend, any job is acceptable (please).*

He finished off as a call came in from the Director of Operations. Mayowa could hear the Director barking at the other end of the line, venting over the long-standing downtime of a major network hub. He patiently waited for Chris to be done with his call.

"No. That, my friend, is the real problem we have," Mayowa countered after the call ended. "Not because we do not know what we want, but because this job does not give us much time to do more for ourselves. There will always be crazy bosses who add to this craziness. When you get home, you are simply exhausted. Or at least, I am." He went on. "I wake up some nights to the incessant sounds of this ringtone, thinking it's a customer calling or a major node is down. I wonder if I am having enough sleep sef[1]. Oh boy, It's not easy[2], jare[3]."

"And it won't get easier," Chris chipped in. "We have to plot our way to the future while working this job, lest we stay here forever." He sighed. "Mayo, it doesn't have to be easy, it just needs to be done."

Transitioning does not come easy. It requires research and dedicated work.

Like Mayowa and Chris, many professionals are tired of being on the same spot and seek to find true fulfillment in their work, which could require transitioning careers. However, transitioning does not come easy. It re-

1 *"Sef": used for emphasis, no actual meaning.*
2 *It is not that simple.*
3 *"Jare": used for emphasis, no actual meaning.*

quires research and dedicated work. Research will help you identify the right role for you, and the knowledge gaps that need to be filled.

Deciding to transition to Tech is a great start - but it requires more than just a decision. Tech is very broad as we have discussed. You need to not only decide which bough of Tech you are interested in, but you need to carefully select what role you would be vying for to get you into the industry. Depending on your interests, you may need to acquire both technical and soft skills.

This chapter will guide you on some factors you need to consider in picking your role in Tech, either as a technical or non-technical professional.

How to Pick the Role That Best Suits You

Oftentimes young people start out their careers thinking they have got it all figured out, but after a couple of months or years, some of them realize that they would rather not pursue a progressive path in their current career. They realize that they would rather be working on something else, somewhere else. Suddenly, the grass on the other side of the lawn looks greener and much more appealing.

Career transitioning requires strategic thinking, time, preparation and planning.

At this stage, many young people opt for a career transition and then jump ship, without carefully thinking it through. However, career transitioning requires strategic thinking, time, preparation and planning. Before transitioning to a new career, you should carefully consider your options. I will share with you some thoughts you could ponder on to guide your career transition decisions, especially in picking the right role for you.

1. **Know your WHY:**

   You must identify and understand your reason for seeking a career transition. The common answers people come up with when seeking a change in career are:

   - *I do not enjoy my job anymore*
   - *I do not have work-life balance on the job*
   - *I need more money*
   - *I seek job fulfillment*
   - *There is no future for me in the company where I work*

   Every research is birthed by a question. Start with the first question that comes to mind

   Identifying your why is key because sometimes your discontent could be a function of a change in job role or department and not necessarily about switching careers across industries. It could also be that you may be feeling lost and you simply need some time off work to reconnect with your inner self. You feeling stuck could

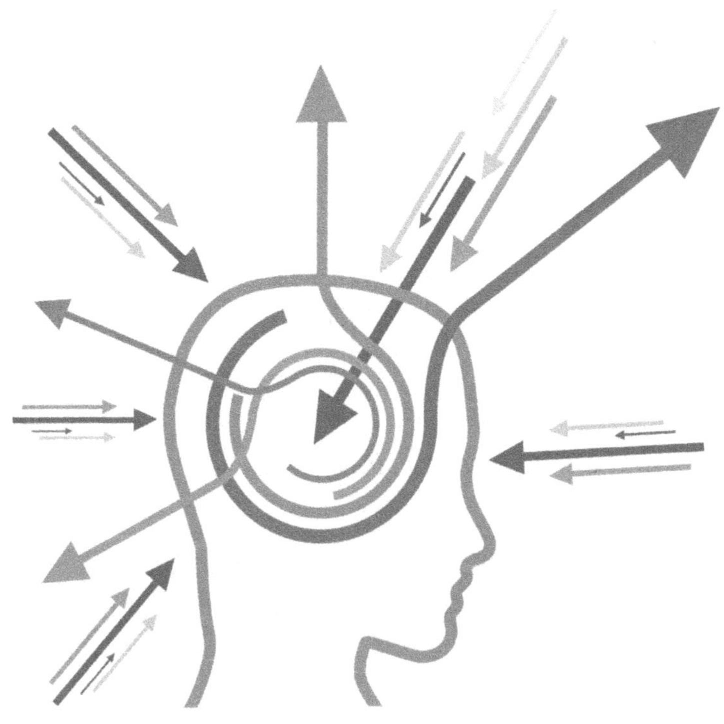

Stimulate your mind!

mean that you need to grow in learning in that same job or industry. The truth is that there is no perfect career per se. These feelings happen to even the most enviably successful people whom we admire - they come and go.

Therefore, never assume that the discontent you feel is solely because your current career no longer serves you. You need to make sure your why is not momentary but is in tangent with your goals and interests. Take some time to answer the following questions for yourself before you decide to transit.

i. *What is the root of the discontent I feel on my current job?*

Could this be the need for a promotion, salary upgrade, recognition, a different role? Or do you truly seek a new expression beyond your current job? Sometimes being on a spot in your career could lead to exasperation, which we all know is a ringing alarm for change. However, career transitioning is too important a decision to be made without the right motives because you just may experience some frustrations along the path of change.

ii. *What is my true goal or vision for my career, and where am I on that course?*

Do not be alarmed if you realize you do not have one; it just means you need to redesign and re-map your career in line with the vision you have for your life. This would be the most critical step before going ahead with switching careers.

iii. *Can my current department, company or industry accommodate what I truly seek to achieve?*

This is why question (ii) is most crucial. Only when you have established your career vision can you truly ascertain if you should or should not make a switch, or better still, get the needed clarity for your career. Sometimes, you may realize that your current company or industry may be able to accommodate the new expressions you seek. You may simply need to seek internal opportunities within your company for a new job role or speak to your superior about your long overdue promotion.

Sometimes, you simply need an uninterrupted vacation to clear your mind, rest or add to your knowledge. Stimulating our minds could be a great boost for our disposition towards work.

iv. *Why am I interested in the Tech industry? Does it support my goals, and can I fit into this industry?*

Interestingly, people have many reasons for seeking to be a part of the Tech industry - one of the usual ones is the great remuneration. However, Tech careers could be quite consuming and demanding. I advise people to have more than money as a motive for switching to Tech, especially if seeking to come in as a technical professional.

v. *Is my decision to transit careers serving my overall career vision?*

Again, motives. Whether directly or indirectly, in the long-run or for a short term, this decision should serve your career goal. You cannot afford to do the Brownian (random) movement with your career because you will end up shooting yourself in the leg at a certain future time T when you least expect.

If you realize after this honest introspection that you truly desire to transition, then you need to move to step 2.

2. **Know your WHAT:**

    Next, you need to know what you seek to transition to in Tech. Even as a technical professional in Tech, you could decide to transition within the industry across boughs, or even across departments within the same bough. What exactly do you really want to do? People have many reasons for seeking a change in career - It could be because of personal interests, or because of perceived accruable benefits. Whatever your reasons, I highly recommend that your interests and anticipated benefits should be at par. This is where you could also get to figure out if you would rather go technical or non-technical in Tech.

    An acquaintance of mine who graduated with a Bachelor's in ICT and got her first job as a technical engineer in the telco space (one of the boughs of Tech), realized after her first year on the job that she would be happier on the business side of Tech and chose to gravitate towards the non-technical side such as vendor management, contract management, et al whilst remaining in the same company. You need to identify which path of Tech you would prefer to pursue.

3. **Investigate:**

   When you have identified the specific field in Tech you would like to transition to, you will need to investigate as much as you can about the industry (if Tech will be a new industry for you), job roles existing in that Tech career, and the job requirements per tier such as qualifications, certifications and time demands. The acquaintance I mentioned above initially had no idea of what being a non-technical professional even seemed like, or what opportunities were available to her - but she began to investigate before getting the answers.

I wish I could guide you step-by-step on researching, but I usually advise people to start anywhere and with your best friend: Google. Every research is birthed by a question. Start with the first question that comes to mind, even if it is not directly giving answers to what you seek the most. Scan through articles, pick the three most interesting or catchy article titles and open them in new browser tabs. Skim through each article, and you will be amazed at how much this exercise will expose and open your mind and how this could lead you to more specific questions that take you closer to the answer(s) you seek. I would also advise you to have a notepad and a pen to write down questions that crop up as you go along. You would find that researching from the general would lead you to the specifics.

Likewise, investigate existing roles, opportunities for growth, salary benefits accruable, career hierarchy and pathways so you could identify exactly where you fit in, what you need and how you could enter the industry if you are transitioning. Identifying and investigating Tech companies you would love to work with is also key. Some websites to assist with this research are LinkedIn, Glassdoor, Vanhack, Indeed and of course Google.

Identify your knowledge gaps and knowledge advantages based on your existing qualifications or experience in line with the industry, company or team you seek to slide into. It is very important that you understand the knowledge gaps required to transit and if they pique your interest. Depending on your desired pathway, you may need to pick up new technical skills and likely some soft skills as well. However, gaining basic yet technical skills in the sector you seek to be a part of will definitely be a booster for you before your interviewers.

SCAN ME

Chapter 4

Prepare.

That day Chris did not go home immediately after work, despite having to work well into the next shift. He instead decided to have a chat with some of his colleagues in other departments to understand the scope of their work. This exposed his mind beyond his initial assumptions, and he was able to identify which job roles pulled at him. More so, leaving the office at the same time with his colleagues who worked the regular 9-5 enhanced his desire for that schedule.

The subsequent weeks saw Chris staying back after his morning shifts or coming in at 8am before his afternoon shifts to learn at the feet of his colleagues: Chima in Sales and Osas in IT. This helped him to better understand the technology that their company sold. Chris also started taking courses in sales, whilst offering to help Chima prepare some sales reports and presentations. Chima was more than happy for the assistance as those were the tasks he least enjoyed.

Sometimes when Chima and his team needed to pitch their solutions to some clients, Chris would ask to tag along. He continued at this for weeks running into months. At some point, he had a breakdown in his health. He was hospitalized for over a week, and his doctor advised him to take things easy. Coincidentally, Mayowa also came visiting that day.

"Guy, you don hear doctor[1]? You need to take things easy man. Cooluo down[2]. With our crazy schedule, you still add extra work for yourself. I don't get you. No be say you dey get extra mula on top sef[3]" Mayowa chided.

"Hmmm! Mayo, I must leave that department o. I cannot be working shifts again. Something has to give," Chris replied.

"So it's until you kill yourself abi[4]? Mayowa asked.

"No pain no gain, bro." Chris answered. "Is there another way?"

"How about applying for jobs and getting one?" Mayowa asked.

"But the jobs we are likely to get are the same kind of jobs. We need extra qualifications or experience for something different. Besides, I am looking to transit to a non-technical

1 *My friend, have you heard the doctor?*
2 *Take a chill pill, calm down.*
3 *You are not even making extra money from the additional effort.*
4 *Right?*

role. That won't drop on my laps." Chris reasoned. "I am simply trying to prepare myself for what I want and that won't come without extra effort. You might want to consider this 'practical' approach yourself." He smiled at his friend.

"Hmmm! How you find the time to juggle it all sef beats my imagination." Mayowa said.

"Time dey, na[1]. It's a matter of restructuring your time, really." Chris responded as he adjusted himself on the hospital bed. Mayowa took that as his cue to leave.

"Las las, we will all get there[2]." Mayowa said as he rose from his seat.

"Amen o." Chris smiled at him. "Thanks for coming to visit bro. I appreciate the company."

"Yeah. Anytime bro. Anytime. Do take care of yourself and I hope to see you at work soon o." Mayowa bid him farewell as he departed.

Two weeks later, Chris was back at work and at Chima's side, learning. The health hiccup could not stop him. He had mentally set a learning goal and a timeline within which to achieve it.

It is possible that by now, you have probably identified what pathway to Tech you intend to follow. If you have not, please

1 *There is still some time.*
2 *In the end, we will get to our wishes.*

do not be pressured. Like I said earlier, this is a major life decision, and it requires all the precaution and time necessary before making such a transition. The important thing is realizing what point you currently are, then beginning your road-mapping by asking questions and sniffing around (just like Chris did) to find what calls out to you. This is also the first preparatory step and could also be a major source of information that could guide you in making the decision to switch lanes.

It is often said that preparations precede opportunities. This is because only the prepared can recognize opportunities. Many people truly desire a change but fail to prepare. This is not because they do not want to prepare but because they are deceived by the false notion that there is no time. There is always time for true priorities. We all create time to eat, bathe, even go to work and probably some form of leisure such as social media, radio or TV. Thus, if transitioning is a top priority, just like Chris, you will make time for it. It will only cost you sacrifice.

Preparation is the legal tender for expectation. Your expectations are valid when you are prepared or are at least in the place of preparation. Like they say, expectation is the mother that gives birth to actualization. How are you preparing for that career switch? Or are you like Mayowa - the "las-las" crew who believe everything falls in place in due time? Well, that is a fallacy - if it were not so, the word stagnation would not exist. That goes to say, everything does not fall in place in due time. Things fall in place because someone moves them into place.

"Chuka, I know I should prepare, duh!" I hear you say.

Yes, but how you prepare also matters. I have seen many people who say they are preparing for an expectation but never seem to apprehend it. What I hope to help you achieve is not just for you to prepare, but for you to successfully transition into your new life in the Tech space.

Envision and Plan

Recall we talked about investigating your career interests? Now use all the information derived from the investigations done above to envision yourself in this new career. See yourself in that career with the required qualifications in your desired company. Your life they say, moves in the direction of your most dominant thought. You become what you see. So, you need to see yourself in that company, on that job, being qualified and competent enough to be recognized and rewarded fittingly.

With that, begin to make plans for your career transition. Ask yourself:

- *What should I learn?*
- *Where do I learn it?*
- *How long would it take me to learn?*
- *How much would it cost me to learn, and do need to start saving for it?*
- *How can I re-organize my schedule to accommo-*

date this new learning priority?

- *How can I build competency in it?*
- *Would I require professional certifications? Which ones?*
- *Which companies am I seeking to work with in this new field of endeavor? Etc.*

You should have a general timeline for your transition, learning plans and pathways. You should also plan to network with industry leaders, experts and contemporaries. I will share more on networking in a subsequent chapter.

Having a volunteer plan is one of the smartest ways to transition into a new industry.

Another key point in making plans for your transition is to have a volunteer plan. Having a volunteer plan is one of the smartest ways to transition into a new industry. Who can you serve in that area of interest so that you could begin to gain viable competencies and relevant skills that will boost your resume and ease your slide into your desired career?

Commence Transition

Transition commences when you start taking active steps towards the change you desire. You have made your plans; it is now time to act. Let me also advise that you do not have to wait to create a perfect plan before you begin to take action. You will only fall into the hands of the greatest thief ever – **procrastination**. Start with what you have and where you are. You could always fine-tune, adjust and improve your plan as you go. Active steps could be baby steps, but here is what your steps should look like:

1. **Learn:**

    You must have identified knowledge gaps that you need to close. So, start learning what you need to learn and build relevant skills based on the job requirements for the already identified job role you seek to transition to.

    Ask Chuka

    You: *Hello Chuka, what and where can I learn from?*

    Chuka: *Hi reader, there are some great learning platforms that offer programs in your chosen field that you could enroll in, but I will offer counsel for both sides of Tech.*

    If you are looking to go into a technical role, platforms like Udemy, Udacity, Univelcity, Alt-School, Decagon, Coursera, Code Academy, edX, Google Qwiklabs, Google Codelabs, LinkedIn Learning, FutureLearn, Khan Academy and some others can help you on your journey to acquiring the relevant knowledge and skills. These platforms offer full course curricula that cover everything you need to learn, and at quite affordable prices too. All you need are your laptop, access to the Internet, your commitment and definitely, rapt attention.

You may hit some bottlenecks while learning, but not to worry, there are also great lessons on YouTube to help clarify any areas where you may be grappling with understanding. Simply type in the subject keywords and you will be amazed at how many tutorials are available to further ease your learning.

For non-technical roles, the platforms listed in the technical roles section above could also be useful sources to learn and prepare. However, you may want to consider other sources such as short programs at universities, educational or professional institutions to further prepare you and boost your profile.

"But I already have the requisite knowledge for the role I seek to transition to," you may counter. I hear you, but I would advise you to make certain of this, and if you are, then the next point could be your starting point.

2. **Certify:**

Get some good global certifications. Whether or not you have the skills or experience thus far, certifications in Tech work like an access code. They would land you an audience and possibly some interviews. The rest is up to you - to sell yourself and your competencies. Certifications are such a great boost because they make up for your lack of experience, especially if you are a new entry to the Technology industry.

Because the Tech space is such a fast-paced industry with strong technical know-how required, employers often love to recruit people who are already competent at the role. So, while you may be a new entrant, having a professional certification tells the employer that you are competent enough to be hired.

3. **Connect:**

Breaking into a new industry can be tough especially when you are unfamiliar with it. It is necessary to familiarize yourself with the companies in these industries. Begin to actively follow companies you are interested in and be familiar with their public activities as well as general industry activities. You should also connect with some industry leaders to seek their mentorship and guidance as you transition. LinkedIn is a good place to start.

4. **Volunteer:**

Taking a cue from Chris, you do not have to wait for a job before you transition. You could commence your transition plan with your current organization. Is there a department that is currently working that job function? Volunteer with that department and find a way to be relevant, not just seek to learn. However, you do not have to limit yourself to that if it is not possible. You could also seek a mentor in that industry and preferably a leader in your desired role to connect and volunteer with, within or outside your organization. A mentor is also sure to give you some form of guidance and could possibly throw some opportunities your way. Alternatively, you may also choose to seek relevant projects that give room for volunteer work amongst other ideas you could think of while seeking to volunteer. Volunteer to serve in projects or events organized by companies in your proposed industry. Seek internships as well. While there may be little financial reward and lots of sacrifice from your end, do not be distracted by all of that. See it for what it is, a great opportunity to improve your skills, boost your confidence, gain some good relevant experience, learn from your mistakes without grave consequences, and of course gain some visibility and a wider network. All of these will garnish your resume and yield you greater returns in the future.

5. **Package:**

While you may not have all the relevant experience required for the role you seek to transition to, you just may have similar experiences you have gained on your career journey.

You will need to put together your resume for interviews. While you may not have all the relevant experience required for the role you seek to transition to, you just may have similar experiences you have gained on your career journey. So, you will need to tweak your resume. Identify areas where your work experience may be closely related to functions in the new job role you desire and emphasize on them. Also, update it with the industry-relevant skills and certifications you may have gained in your preparatory process for transition. You could also seek help online or use professional services to help put a resume together for you if you find it somewhat challenging to adapt your experience(s) to the required role. Some self-service platforms include resume.io and careerlyft.com. More importantly, always keep your resume handy and updated.

6. **Apply:**

   Now to the crux of the matter. You need to seek and apply for jobs in that role and prepare for interviews consistently. Do not be discouraged because you are likely to hit some road bumps on this new venture. Rejections may likely come your way, but you need to be resilient and learn from them. What could you have done better? Is there a better way to answer questions or package your resume? Do you need to further upskill? Whatever the case or lessons may be, keep up with the applications until you land your dream job. Notwithstanding, even if you do not land your dream job yet, I often advise people to sometimes take up opportunities to learn and grow your competency in your desired role even if the pay package is not what you are hoping for.

Pace Yourself

All of these may seem like a lot, but like it is often said, "A journey of a thousand miles begins with the first step". Make a plan and take a step, even if it is a baby step. Baby steps are better than no steps at all. More importantly, pace yourself. Transitioning to a new career is a life-changing decision that should never be rushed into. Give it time, but not too much time. Set timelines for achieving each transition goal. Set timelines for envisioning and planning, learning, getting certifications, volunteering, etc. I would recommend between 6-24 months depending on the skills and experience you may need to break through.

Pace yourself financially.

Also, you should pace yourself financially. Quite a number of people who seek to transition often resign their jobs and plunge right into trying to get a new one. Except you have a strong financial backing, this is truly unwise. More often than not, they end up caving in to the difficulties and uncertainties that come with transitioning. Pace yourself financially. Start saving for when you may need to resign so you can intern or volunteer or probably even wholly commit yourself to just learning as the case may require. Having some money saved up to sustain you for several months is a great plan. So, before you put in your papers on your current job whilst seeking for greener pastures, pace yourself.

Finally, I leave you with these wise words:

"

Unfortunately, there seems to be far more opportunity out there than ability... We should remember that good fortune often happens when opportunity meets with preparation.

Thomas A. Edison

SCAN ME

Chapter 5

Finding a Job.

Ask Chuka

You: *All these things are great Chuka, but I am literally a novice in this Tech industry, without contacts or connections, and with little or no experience. How can I truly land a job in Tech?*

Me: *That is what I am about to unveil to you in this chapter, so please follow me closely. Get a notepad and a pen to note relevant action points or take-aways.*

Do you remember how you landed your first job, especially if that first job was in Nigeria?

Many people have different experiences. Some landed their

first jobs as they were graduating, but a common experience in the Nigerian labor market is that of persistence amidst rejections. As fresh graduates, we all had in mind the big names and brands we hoped to work with, but just a fraction of us actually got into those juicy companies or corporations. The competition to get in is usually fierce as you probably had to vie for a slot amongst thousands of other interested, competent parties. Besides, there is a limit to the number of intakes these companies can accept.

Then come the possible rejections. Some of you reading this may have a similar story. It took several refusals, rejections and even close calls that eventually did not pull through before you landed your first job. But one thing was clear, it required persistence and patience.

I have taken you back to the past because transitioning to Tech, especially as a novice, is very much similar to seeking out your first job as a fresh graduate, regardless of the years of work experience you may have gathered. While you could land a job immediately, you also must recognize that if you do not, it does not mean you are on the wrong path, neither is it a futile effort. With persistence, it will surely pay off.

Transitioning to Tech, especially as a novice, is very much similar to seeking out your first job as a fresh graduate

Mission Mind Priming completed! Let's get practical now.

So you have picked those fancy skills and done some interning or volunteer work, yet you have not landed a job to kick-start your career in Tech officially. These six tips are sure to help you along your journey.

1. **Search online:**

    Search out companies in Tech and get to know about them. Understand their business, their market, their customers, their unique offerings and solutions. The beautiful thing about online search is that similar companies will also pop up as you search, broadening your scope. You probably would love to work for Google, Facebook, Oracle, Andela, Apple, Microsoft, Paystack, Flutterwave and the other global brands - but so does everyone else. While you may be lucky enough to land a job at any of these big names, there are so many companies in the Tech space besides these ones.

    These big companies have vendor companies and even sub-contracting companies that serve them. This means searching out these companies to even know them is a good first step. That way you will have a broader area to search for vacancies you could apply to. There are online sites as well that could help you with available vacancies in these companies. That way you do not have to be hopping from website to website, looking for vacancies. LinkedIn Jobs, Stack Overflow Jobs, Remote.com and Vanhack.com are some websites that will definitely help your search.

2. **Start with your "connect":**

    You may know someone who is working in Tech already,

or someone that knows someone working in Tech. Start from there. Indicate your interest to that person and ask the person for advice on companies to apply to for known vacancies. You could even tell the person the kind of opportunities you seek and ask them to kindly put out an ear for you if any opportunity comes up. More assertive people may even ask those people to recommend them to recruiters they know or put in a strong word for them.

You will be amazed the benefits this little common step could accrue. Do not accept "No" easily. Seek out different ways for that person you know in Tech to help you. Squeeze out that juice, no matter how dry. Whatever you decide, do not just wait for a job to drop on your lap. Go find that job in every legal and moral way possible.

3. **Use social media:**

   These days a lot of recruiters refer to social media to profile potential recruits. You want to make sure your social media is updated and appropriate. For instance, many of them will view your LinkedIn profile. How aptly put together is it? Are your achievements and work experiences listed? And your picture! Is your picture professionally appropriate? LinkedIn is not like other social media platforms such as Facebook, Instagram or Twitter. They may refer to the others but will likely start from LinkedIn. You should also add a touch of your professional side to all your social media platforms. You should be intentional about the image you project of yourself, your manner of speech and tone. Also, your profile description should be an eye-catcher.

With LinkedIn, you could reach out to recruiters to pitch yourself for posted jobs. Also, you may want to get the best out of professional social spaces such as LinkedIn by paying for the Premium service if you can afford it. On the other hand, if you cannot, LinkedIn offers all its users a 30-day free trial for its Premium package which you can maximize.

You can also specify your job search titles and location so that you could get notifications on job alerts that match your interests. This will bring you opportunities from companies you have never heard of in your desired industry.

4. Network:

I have decided to write this as my next chapter because I want to help you squeeze the juice out of this orange to the max. However, to whet your appetite, I would say this about networking. Are you really networking, or do you assume that you are? Do not be caught up in your competencies. Package yourself. Market yourself and sell yourself to a worthy bidder. You do not have to be most proficient at networking, but you certainly need to network.

Is that a "Chuka tell me more?" Then see you in the next chapter but finish this chapter still. Lol.

5. Apply! Apply!! Apply!!!

How would you get a job if you do not apply? Rejection is the food of champions. Do not get discouraged and throw in the towel. Good things sometimes take time to come. So never give up. Keep applying. Do not be selective about companies as well. Sometimes you may start small but with vision, hard work and consistency, you will end up with the recognition and reward you deserve. Besides, more often than not, you get more experience and exposure in the smaller companies. So give them a shot as well.

You may start small but with vision, hard work and consistency, you will end up with the recognition and reward you deserve.

Your goal at this stage is to get a job that can validate your knowledge and give you referable experience that will boost your resume. Remember, you are moving to a new level with your transition, but you are starting from the scratch again at this new level. So, regain the hungry and lusty mindset for knowledge that you once had as a fresh graduate. That way you will learn much more and grow faster.

6. Follow up:

Do not just send in applications and relax. You need to follow up. I know that the automated way recruitments are done these days leaves little room for follow up because the mails are automated responses. However, for the applications you directly send to a recruiter's mail, follow up. After interviews, seek out from your interviewers how to follow up. They may not give you a contact though, but you may want to be mindful during the introductions to recall their names. Search them out via LinkedIn and connect with them after the interview. Do not be a buzzer please. But you may follow up with them after some time if you have not heard anything. Remember those contacts you asked to keep an eye or ear out for you or to recommend you? Follow up with them. Follow up on all opportunities and leads via emails, phone calls or direct messages on the social space. Definitive responses can help you make better use of your time and burst any bubbles of false hopes from growing bigger with illusion.

Sabali (Patience)

Getting a job is a process that requires time. You will need to be patient. Neither rush it nor give up too early. It will come. Like it is often said, *"Patience is not the ability to wait, but the ability to keep a good attitude while waiting."* So how are you waiting?

Be positive.

When you are full of positivity, your vision and focus get clearer. You are able to keep moving towards your target (a career in Tech). I know it could get frustrating charting a new course, and uncertainties do not make it easier on our minds. However, if you truly are resolute about your decision to transition, then adopt a no turning back approach and stick it out. Being positive is what will help you wait out the process. Never internalize the rejections that may come your way. Accept them, learn from them and keep pressing on with hope. One "Yes" is all you need to shut off the noise and sting of a thousand "No's". So keep your head up, keep your heart beating, and your hands busy with learning.

Keep your head up, keep your heart beating, and your hands busy with learning.

Stay active.

Until you reach your destination, stay on course. Keep driving, keep navigating; be consistent with your job search. For every "No" you get, swing back by sending out two extra applications. You may even send unsolicited applications to recruiters. Reach out to hiring agencies if need be.

Stay motivated.

Do not stop learning. Try out more projects, seek out more volunteer opportunities, keep networking, keep exposing your mind to the business of the Tech industry. These will help you stay motivated and grow your confidence, competencies and connections which you will find highly invaluable in your career journey. Always seek out ways to maintain your relevance and goodness will knock on your door soon enough.

Be expectant.

Be fervent in your expectations. When you are expectant, you will see the many faces of opportunities. It may not come as expected (a job). It may come as an opportunity to be a partner or as a business. What you do not expect, you may miss. Therefore, you need to be open-minded as well.

Sabali!

SCAN ME

Chapter 6

Networking.

Networking is about making new relationships for the purpose of mutual benefits. It is purposeful and deliberate. This is a key tool for career success in the 21st century as the level of your network determines the kind of opportunities that come to you, which in turn determine your growth rate. Like Porter Gale said, “Your network is your net worth”.

Most career professionals are indeed familiar with networking and its importance - but many of them do not utilize this tool on their career journey. This I have narrowed down to three core reasons:

1. **Lack of skill:**

   Networking requires skills and strategy. Quite a number of people lack both, or are unaware that they are needed in order to effectively network.

2. **Misconception of networking:**

   In Nigeria for instance, there is a tendency for employees who put themselves out there to be labeled as being grandstanding. In other words, some people confuse wanting to network as being pretentious, self-seeking and overly assertive - so they would rather not.

3. **Ignorance:**

   Though many have heard of networking and its benefits, there are some others who still believe that if they work hard and well enough, they will eventually get the recognition and/or the promotion they deserve. While that sounds noble, it is important to wake up and smell the coffee. Times have changed, and you need to adapt to the times you currently live in. Life does not give you what you deserve, but what you demand.

People who are less competent tend to be the superiors of the more competent hands these days. Their networks give them the opportunities to get ahead.

   People who are less competent tend to be the superiors of the more competent hands these days. Their networks give them the opportunities to get ahead. While opportunities are still baking in the oven, the smart, less competent ones seize them - and then use the more competent hands who are still stuck in 1950 to ensure quality performance and excellent service delivery.

“

Networking is not about just connecting people. It’s about connecting people with people, people with ideas, and people with opportunities.

Michele Jennae
Author of The Connectworker

As a career professional, you do not want to have the cold leftovers of a long-abandoned box of pizza. You need to network. Below, I share with you some keys for scaling your career using networking as a tool.

1. **Network:**

   Firstly, the way to network, is to network. You cannot do that by being an island. To network, you need to go to where people of like minds are. Go for events, especially in your industry or in areas of your interest. If you do not know what events are happening, Google is a good place to start. You could search Google for events happening in your city within a specified time frame. There are some websites such as Eventbrite, Eventful, 10times or ViewLagos that also offer you this information, amongst others. Please note that while these are in Nigeria, your domain will definitely have sites like these. Just start with Google.

2. **Investigate:**

   Networking is not about merely socializing. According to Diane Helbig, "Networking is an investment in your business. It takes time, and when done correctly, can yield great results for years to come." If networking were an investment, you would want to first investigate that investment to know what benefits will accrue to you in the long run, as well as just how to invest. In the same vein, do some research about the event, speakers, organizers etc to find useful information that could help you connect with people and make meaningful contributions.

3. **Prepare:**

   You need to prepare yourself thoroughly, mentally and emotionally (this is not the time to give in to the pain of a heartbreak). Prepare physically - your outfit, your appearance, and even your gait should be graced with style. Prepare your introductions. You will be needing to pitch yourself without being obvious. You need to prepare so as not to be too cliché. You also may not want to be too scripted - people will lose interest. No one wants to talk to a robot. Then, have a networking plan. The subsequent points will elucidate on this.

4. **Scan:**

   Much like we have established, networking must be deliberate. When you get to the venue, do not get carried away with the ongoing activities. Scan the room to know people you would like to meet so that as soon as it is time for recess, you can make your move. Most people like to go for the guest speakers. While that is great, your face is likely to be lost in the crowd of faces the speakers would meet that day, except you have prepared and mastered a good pitch.

   You may also want to be the first they meet or one of the last they meet (if you insist on meeting the guest speaker) as you would be more memorable that way. Nonetheless, there are so many others you could network with aside from the guest speakers. Consider all those seated on the front row as well. Look out for industry leaders or influencer's also present. Scan your contemporaries and even some younger ones and decide who you would love to connect with - before it is time for the tea break.

5. **Contribute:**

   Ask questions when opportunities are given for questions. Even if you do not have any, you could just ask a question (not an unnecessary one though, please) or make a contributory remark depending on your confidence level or assertion. Standing up makes you a face in the crowd and people are likely to remember your face and be receptive to you when you approach them. Chances are that you may even be approached by some people during the tea break for that act of intentional boldness (good stuff!).

6. **Converse:**

   Ditch excuses such as shyness and being introverted. Do not wait for anyone to come meet you. Be proactive about it. As the tea break commences, people would like to pair up or cluster together. I recommend you join clusters rather than pairs. You will feel less of an intruder that way. You can also pick up on the conversation which is more likely to be general as compared to a private conversation between a pair. Also, never lose sight of your goal. Do not get carried away when in a cluster. Avoid forming a "clique". You are there to meet as many people as you can. Involve yourself in conversations

Find mentorship and guidance from someone in your ecosystem

“

You need to assert yourself at the forefront of people’s busy minds. Follow up!

Chuka Ofili

with the aim to casually share at some point what your strengths are - do not fail to do this!

Ask Chuka

You: *But Chuka, I do not have strengths yet.*

Me: *Sell your strongest professional point and link it to Tech. Round it off with a request for guidance or mentorship or opportunities from the strongest person(s) in the group you perceive may be able to assist you on this journey.*

7. **Again, follow up:**

   This is so vital - yet many people fail to do this. Everyone is busy. Even the idle person is busy being idle. You need to assert yourself at the forefront of people's busy minds. Networking is not just about meeting people, it is about following up, creating relevance and building relationships with mutual benefits. You must plan for the number of people you can comfortably follow up with while networking. You could meet thirty people for instance (that could be a bit much, lol) but know the top three you intend to follow up with after that day - and get in touch with them within 72 hours.

   For those who are in IT and would love to know some professional events to look out for, here is a list of some which happen regularly. Google will still be your best friend to know exact dates and time.

- Google Developer Groups.
  (Happens in several locations across Africa)
- Google Cloud Next
- AWS Re-Invent
- TechCircle

SCAN ME

Chapter 7

Impostor Syndrome.

Congratulations!

You have finally landed your first job in Tech. Despite the time you have spent preparing and getting relevant skills and certifications, it really is a new industry with different processes and systems, all likely to be completely foreign to you. Coupled with the pressure of being the new face and with you probably also being on probation, the pressure to blend in, perform and prove yourself could be very daunting regardless of whether you are in a technical or non-technical role.

There is also the phase where your notes and learnings seem to have flown out of your brain while working on this new job. You recognize the scope of the matter: you are familiar with that knowledge area involved but somehow you cannot seem to prove by solving the mathematical equation that makes **2 + 2 = 5** (on a lighter note, this is valid mathematically).

Then you begin to feel out of place, overwhelmed and incapable. Everything suddenly seems new to you and it feels like your brain has become spongy with many holes in it. You may even begin to ask yourself what the heck you were thinking transitioning and how you probably should have remained in your comfort zone.

This is like the proverbial Peter who was eager to walk on water - which he did until he realized *"Crap! I am walking on water for real. That is not possible!"* - And at that moment began to drown (lol). Have I just described you? Perhaps, you are wondering what Chuka is on about. Then lucky you, because this will prepare you for this highly probable experience when this happens.

All of these are normal, but beyond being normal, you should know that these are symptoms of *Impostor Syndrome.* You need to know what you are dealing with to deal with it appropriately and terminally. Those feelings of self-doubt, of not being good enough, of feeling unworthy of your accomplishments because you were *"just lucky"*, of feeling like you will run into a challenge on the job that will expose you as a fraud who got a role that he/she was not qualified for all have one name – *Impostor Syndrome.*

Breathe.

This happens to the best of us, and I am dedicating this chapter to prepare you to confront this so you do not go self-sabotaging on your new job. Understand that book knowledge is very different from practical knowledge. All that academia prepares you for is to face the challenges that come. Selah. This means that despite your knowledge and preparation, you will still face situations that will challenge that knowledge or preparation. Accepting this will help you brace up for those moments. Even the experienced hands in Tech still encounter incidents that confound them.

Attempt.

Do not let that situation get the better of you. Do not get overwhelmed, negative or worse still, throw in the towel. Challenges come to not only test your knowledge but to validate it, and most importantly, to grow your knowledge base. Adopt this mindset. Put your knowledge to the test. Embrace the challenge by attempting the task. Do not just say you do not know it, try. You most likely already have the requisite knowledge but they are like scattered puzzle pieces that you need to piece together. Break down the task into components and determine what will be required to develop each component. As you do so, look out for knowledge gaps that will hinder your work and seek to close them.

Learn.

See the challenge as an opportunity to learn and hone your skills. Learning means to reach out to someone to teach or clarify for you the problem area and its solution. This you could do by reaching out to a colleague to guide you through

“

You miss 100 percent of the shots you never take.

Wayne Gretzky

the gray areas. It does not make you any less competent or qualified. It shows your great team spirit instead. However, I would advise you first consult your best friends – Google and YouTube. Seek to learn first from available sources and you will find a ton of materials there for that problem area. If you are still stuck after trying to research, then you could always consult a colleague. It is important to point out that in doing this, you must take heed to yourself so you avoid being spoon-fed. Ask pointed and specific questions and do not pass the ball. Do ensure that you take notes as much as possible, where applicable.

For instance,

> "Hi Chuka, could you please help me roll back? I don't know how to do it."

A more appropriate way to ask is to say:

> "Hi Chuka, I need to roll back, could you guide me through so I could implement it?"

The first approach shows that you are not interested in learning but in passing off the responsibility to another. Your disposition with this approach will probably also be similar - a way to get the task off your plate in the easiest and fastest way possible - and you will not have an open mind to receive learning. The second approach on the other hand is more professional, serious-minded and shows a willingness to learn and improve yourself. You may get away with the first approach at first but after a couple of times your colleagues may refuse you and even begin to talk about your "laziness".

Teamwork.

Seek out opportunities to work on projects that require teamwork. This way your skills will be harnessed, and you will learn a whole lot more. In a team, you could safely make mistakes and someone most likely will be there to clarify and correct you early enough. Working in a team will also increase your exposure career-wise and give you a more holistic view of your organization's services as you also begin to find a niche for yourself. When you work in a team, you build connections as well.

Refer to previous learning.

Proficiency is the assembly of all the building blocks of knowledge you have in a particular area into a systematic structure. You will find that on the job, you will need to refer to previous learnings to resolve some issues you may encounter. However, remember to explore and expand that knowledge you have. Do not work strictly by the book.

Proficiency is the assembly of all the building blocks of knowledge you have in a particular area into a systematic structure.

Remember as a child when you first started learning? All you knew were the numbers 1-10. But as you explored the world of academia, you learnt the different ways you could use numbers such as addition, subtraction, multiplication, division, and the entire BODMAS suite. Just when you thought that was all the complication that existed, logarithms and exponentials, differentiation and integration, permutations and combinations and the worst of all, the 10th planet of the earth – statistics, came to play, further complicating the world of BODMAS you had managed to finally crack. Lol. Guess what? You learned them all, at least to a certain extent.

You will likely be faced with new worlds and a new planet (much like statistics) on the job and that's okay. Never forget that they are rooted in the foundation of "1,2,3..." and "BODMAS" – the basics. Always refer to the basics, explore and then expand.

Hone new knowledge.

This is pretty much like the new world of logarithms, exponentials, statistics, etc. You need to not just learn these similitudes as it relates to your field, but also hone the knowledge. Recall as a student how you always had this particular module you could smash even in your slumber? You need to move beyond the numerals.

As you explore the new world (by now I believe you understand I mean exploration of new knowledge, and not Christopher Columbus' *Exploration of the Western world*), some things will call out to you. You need to hone them and build on them as strengths. Some other things will bark at you - your weaknesses - you need to learn beyond the basics and stretch to at least elementary level if not intermediate while the rest may wink at you, signaling opportunities. These opportunities may likely come in the form of knowledge in line

“

I like my job because it involves learning. I like being around smart people who are trying to figure out new things. I like the fact that if people really try they can figure out how to invent things that actually have an impact.

Bill Gates

with upcoming trends in which you need to build competencies so as to stay ahead of the game and sustain your relevance.

You may be a newbie but daily improve on yourself. Never let yourself remain in a place of ineptitude. If you have a task you do not know how to go about, learn and hone that knowledge so well and deep that you will be able to recognize "Arya Stark" even with her many changing faces. We will delve deeper into this in the next chapter. *Don't drop this book yet, you're almost done.*

Win to build confidence.

When you attempt what you have never done and succeed at doing it, the gratification (if not the thrill) is such a pleasure. So, look out for ways to win. As a newbie or rookie, if I may prod you a bit (lol), each uphill task you complete, should be celebrated by you (internally at least). Use it to build your confidence and doggedness. Always remind yourself that if you could execute xyz task which you thought will totally screw you over, then you can most certainly handle this new **բարդ** task. I just heard you. *"Chuka, what is that?"* You say.

Focus on overcoming that which seeks to block you. More wins help your confidence levels and boost your morale

Hahaha, that is how some tasks are, seemingly gibberish but it means something just as "**բարդ**" (pronounced /bɑːrd/) means complex in Armenian. With time, nothing will faze you again and you will become an experienced and trusted hand on the job. Again, focus on overcoming that which seeks to block you. More wins help your confidence levels and boost your morale even if it takes you a longer time to accomplish it than the average person.

Fraternize.

I know it's both tempting and overwhelming at the same time, but refuse to be a face in the crowd. Meet and mix with your colleagues. Do not wait for people to come over to you, be assertive and go over to them. This way, you will also feel a lot less like an impostor. Claim your environment, make it home. After all, you will spend most of your day there. When your work environment becomes home, you will see that you will feel more at ease psychologically. Being the new guy often gives you that uneasy feeling that all eyes are on you and there is a tendency to be a wallflower. That is just your mind. Yes, all eyes are on you, but most people are more eager to know who you are than see if you are good at what you do because man, at his core, is more social than competitive naturally.

In other great news, this would mean that they will also be more than glad to help you settle into your new role. Besides, what is work without having a great work buddy or two? So, get into it. You are not an impostor. Meet and mix. You will be just fine.

SCAN ME

Chapter 8

Time Management.

Being a new entrant in the workplace can be challenging at first but take your eyes off the challenges and consider how far you have come - from having little or no knowledge of Tech to successfully transitioning to Tech. You have done the seemingly impossible (**#applause**), so you can definitely overcome the challenges of your new job. I know it may feel a little bit like you are out of step but learning to manage your time at work will help you improve performance at your tasks.

Time management is one area a lot of people struggle with generally in life. Working on a new job with new challenges in a new industry can make this struggle even more frustrating. It is not uncommon to battle with the tasks assigned to you. Meeting up with deadlines when simply trying to navigate the maze of work further compounds the problem and you begin to really feel like an impostor (refer to Chapter 7). Clarity and speed come with time and practice for some people - it doesn't mean they are not good enough. For some others, it is a walk in the park.

I picture a scenario where you have finally made your move to Tech but are not quite sure if you made the right decision. The workload is heavy (typically, so welcome aboard), you never seem to finish your tasks or even fully understand them often enough, you are working late hours, barely getting any rest, you have a lot more to learn and there is hardly any extra time left for learning. It feels like you have lost your life balance. Does this seem like you? I am not going to console you by telling you that it happens and is a part of life in Tech. Everyone does have to battle with work pressure, and you must find your balance between work and life.

Not having this balance will cause you to work under pressure. In short, work will always come with pressure. Here's why you need to handle pressure on the job:

- *Pressure will lead to stress*
- *Stress will cause you to lose focus of your core objectives*
- *Loss of focus will lead to poor time management*
- *Poor time management will reduce your efficiency at work*
- *Reduced efficiency will yield to lower productivity levels on the job*
- *Diminished productivity over an extended time*

could earn you sanctions on the job

- *All these could lead to a breakdown in your health or even loss of life, family, relationships etc at any point.*

Effectively Managing Your Time

Finding your balance starts with managing your time but you cannot effectively manage your time until you discover how exactly it is you misappropriate your time.

Illusion of time.

The first culprit will be the illusion of time. Most of us often live and work with an illusion that time is more or sometimes less than what it really is. Let me explain this. In a training room once, I saw the trainer engage the participants in a little time experiment. He asked them all to close their eyes, got a timekeeper to mark the time and instructed the entire room to be silent. The game was simple: do a mind count and indicate with a show of your hand when you feel that two minutes have elapsed while the timekeeper records the exact amount of time that you consider to be two minutes.

To my amazement, nobody could tell exactly when it was two minutes. Only one person indicated at a minute and twenty seconds while the rest were beyond two minutes. The person next closest to the mark did so at two minutes and thirty seconds. Others surprisingly did beyond three minutes, and the last person even lasted for five minutes and counting. Quite hilarious - but revealing too.

How we perceive the time we have determines how we manage our schedule. I am not saying this to put you under time pressure. You must avoid that. Nonetheless, you must have the right understanding of the expanse of time to rightly manage it.

In what ways does this affect your task accomplishments? Do you sometimes fall into the illusion that the task to be turned in next week is still a month afar off in your mind?

Punctuality.

This is such a challenge for many but very key to any endeavor. An old quote says, *"Punctuality is the soul of business."* How punctual are you? Do you arrive at work in time to organize and plan your day? Do you get right into your tasks or do you love to settle in (as I see some do) to gist and catch up with your colleagues before getting into the day's work? This is not necessarily bad on its own, but it is imperative to always organize your schedule first.

Punctuality is also beyond arriving early at work - it is a mindset and an approach to life and things. Do you strive to be punctual in turning in your deliverables? Or do you have an *"anytime it is ready"* approach? When you do not place a firm grip on turning in your deliverables within set timelines, you will be relaxed and then fall under pressure when it is requested for. I always advise that you ask for expected timelines for delivery when given a task. Be clear about it and then work to turn it in earlier or on time at the least.

Overload.

I see a lot of people take in more than they can chew on the job. While you may be trying to pass across the impression that you are available and hardworking, what your superiors are really looking for are people who are reliable and dependable. They seek those who not only take up tasks but deliver them with minimal supervision and more importantly, minimal corrections. So, the real question is: just how dependable and reliable are you? How accurate is the quality of your work? That being said, you need to know when your plate is full and then say “No”. Your plate being full does not have to do with the number of tasks you have pending only, but everything to do both with the tasks you have pending and the time-frame needed to work on and deliver them. You must always place this in line with your allotted time for work. You should be firm and strict to the best of your ability.

Also, while work tends to extend beyond the official hours, you should set a reasonable allowance on how much overtime you put in. All these would guide you in knowing when you are overloaded. As a newbie, you likely would have to put in overtime. In fact, working in Tech requires you to put in overtime, but you must also know when to draw the line and strike a balance.

Poor planning.

This is so key and a great follow up to work overload above. Like an acquaintance of mine says, “If you fail to plan your work, schedule your plan and then prioritize your schedule, you are prone to being overloaded and working under pressure." Many people work on the go and suffer a great deal of disorganization on their jobs. Do not blame your bosses - they will always give you more work. Plan your work and learn to stick to the plan within flexible limits of *workload, delivery timelines and working hours.*

Speaking practically, you may decide to put in extra 10-15 hours of overtime in a 40-hour work week for example. This will help you estimate feasible delivery times for assignments as well as the amount of work to take on. Planning is essential so that you could be more productive and then have some form of balance.

Here are some tips for planning your work.

- **Make a work list:**
  Prioritize what you have to do, and what should be on the list.
- **Add level of effort:**
  For each task, estimate and add how long it will take to get it done.
- **Account for research effort:**
  There will be tasks where you know nothing about the subject matter. Always add a buffer of time for research work.

Time wasters.

There are a lot of distractions that could steal your time at work. Phones, Internet, social media, text messages, meetings, office banter etc. The solutions to all of these are discipline and planning. I like to put my phone with its face down while in focus mode. That way, I do not get sucked into social media or get distracted by phone calls or text messages which can be seemingly fun but are huge time wasters, because we are at work to work. All that time spent will not be regained while the work will be pending and piling as more tasks are added to you. Notwithstanding, I realize that we live in a connected world, so to keep up, I occasionally take several fifteen to thirty-minute breaks throughout the workday.

This formula is not cast in stone - you could always plan what works for you.

Sometimes as well, meetings could be huge time wasters as valuable work hours are used for these meetings which usually have lots of back and forth. Unfortunately, they are necessary evils we must face because they are important to the operations of the business. I prefer to schedule meetings (especially when within my control) towards the end of the day preferably, or at the start of the day. This will free up the rest of my day for me to be focused and effective at my tasks.

Automating processes and tasks are huge ways to free up your time.

There is a lot of loop time wasted as a result of re-work. We often spend time doing the same things over and over again. My key to this is to Automate. I have a personal rule: If I've done something twice, I won't do it a third time if I can automate it. Automating processes and tasks are huge ways to free up your time. For instance, how can running a program via Python fast-track your data reporting? What other productivity tools could ease your work and save you some time?

You could suggest it to your team lead so that it could be adopted by the team or company. Even if they do not adopt your suggestions, you could personally invest in such tools to ease your burden. Remember, your work-life balance is entirely up to you and should be a priority. Investing in time-saving tools or applications would save you some time and improve your efficiency and productivity.

Inadequate recreation and rest.

Do not be so lost in work every day that you never take breaks. Your brain needs some rest and recreation to keep functioning at top-notch. Many times I see people who take their lunch breaks to simply just eat which they do in the least amount of time before they get right back to work. While this may look noble, this is not the best. It is not just about observing your break time, it is about what you do while on break. You need to ensure that you rest or recreate during your break time. It should not just be about eating. You should come back refreshed and reinvigorated to handle the rest of your tasks. Inadequate rest and recreation will cause your brain to be cluttered, your mind to be unbalanced and in general reduce your coordination and effectiveness over time.

Inadequate rest and recreation will cause your brain to be cluttered, your mind to be unbalanced and in general reduce your coordination and effectiveness over time.

Adequate rest and recreation will boost your mental balance. This goes beyond your break time to even what you do when you are done from work: your weekends, your leave off work etc. Endeavor to refresh your brain as often as possible.

There are many things you could do to recreate, some of them even while working on your seat such as indulging in some comedy to laugh, seeking new inspiration, practicing some meditation, breathing exercises, traveling, hanging out with friends or colleagues, networking, going on dates etc.

What are the other hindrances to you effectively managing your time? When you are more observant of the hindrances to your effective time management, you can then begin to improve on them and reap the balance thereof. This balance is vital to you being able to improve on the job as you will be able to create time to take up new learning opportunities which are vital to your career progression.

Continuous Self-improvement

In today's world, as technology is constantly evolving so are its uses, cases and applications. With this evolution comes upgrades in the software and applications that power them, as well as new systems and processes.

As more companies adopt these changes, new job functions and roles will emerge which could over time cause current job functions to be obsolete. Case in point: the role of cloud architect or cloud engineer is one which did not exist 10 years ago but is a sought-after role today.

Consequently, there are jobs which do not exist today which will in the future as a result of evolving technology. This is why you cannot be too caught up working and not growing Growth is evolution and you must evolve while on the job. It is necessary to keep improving on yourself to stay current with the evolving times.

I saw you coming with the "no time" excuse, and that is why I did an entire section on time management!

One major excuse I often hear is that of no time. Remember the earlier story of Chris who was determined to transition careers? Despite his crazy schedule, he made time out of no time and of course he volunteered for Osas as a way to place his foot in. Let me tell you the rest of the story in a nutshell.

Chris performed so well that Osas rooted for him to be a part of his team when the vacancy arose. Within two years after Chris officially became a part of the Sales team, Osas got a new job in a different company and recommended Chris as his successor. Chris was so good at his job that it was basically a landslide into the big shoes of Osas. While there, he performed very well and within a not-so-distant time, Osas once again recommended him to join his team in the new company. Yes, for a greater role with much higher pay.

Mayo on the other hand was later moved to another slightly higher-paying cadre, still as a network analyst within the same company. He had gained no additional skills. Fast-forward five years to this present day: Mayo has retraced his steps and is currently making the sacrifice to learn and volunteer for his desired job, despite his busy schedule. To get to this point he had to identify what he wanted as a career which was to become a cloud architect. He now takes daily steps towards achieving this goal.

Never compromise your future for todays work. Take the pains and make the sacrifice to improve yourself

Like they say, if you fail to plan, you plan to fail. Time is what you make of it and what you call it. Never compromise your future for today's work. Take the pains and make the sacrifice to improve yourself like Chris did. No matter where you are on your journey or how much time you may have lost because you were moving at no speed or snail speed, like the Nigerian adage says, ***"Your morning starts when you wake up."***

Two major hindrances to self-improvement as I have observed and vividly depicted in the story of Chris and Mayowa are the misconception of "no time" and the failure to visualize and envision career goals and direction.

Practical Tips for Continuous Improvement in Your Tech Career

1. Stay updated:

To remain relevant, you must stay up to date with current trends and future targets. See how your work, skills and interests can position you for relevance in the future. This will require a lot of research, reading, forecasting and engaging to get relevant information you may choose to act upon or use to guide you. *Inquisition* and *curiosity* will be your best friends here. They will keep you hungry for information.

2. **Add knowledge:**

   There will always be knowledge gaps. You need to add current trends and skills to your existing knowledge. Create time *(again the need for proper time management - I can hear you say "Thank you Chuka")* to enroll in classes or workshops to learn more about your interests. Learn the latest trends, update your skill sets etc.

3. **Practice:**

   All you learn will be useless to you if you never practice them. Practice as you learn. Avoid waiting for the imaginary time ***'T'***. Practice to build perfection.

4. **Be open:**

   The Tech ecosystem is very dynamic, and only dynamic individuals can thrive in Tech. You cannot afford to be set in your ways. Be open to change. Embrace the new. Be quick to adapt and be wise to adopt. Join the movement, move along with the waves. Also be swift to learn other ways to kill your target bird. Tech is powered by speed. You will be amazed at what you can learn from your colleagues and even subordinates if you are open and willing. You may learn more enhanced and faster ways to eliminate inefficiency.

Just like software versions always have upgrades, you must upgrade yourself in Tech to have a fulfilling and rewarding career in Tech.

SCAN ME

Chapter 9

Give Back.

I picture you successfully transitioning careers and enjoying a fruitful and rewarding time in your Tech career. One thing is essential though - you need to give back.

Once you are grounded in your new role, consider giving back to the Tech ecosystem. Look around you. Who can you help? As you come across different people in different stages of their careers, give back. You were once an intending or new entrant and you needed a lot of assistance. So freely help someone else who might be in the same spot you found yourself when you began your journey to transition. Give them quality advice and mentorship.

It is so important to give back. That is the way to sustain the ecosystem. No system can survive, much less thrive, if all it experiences are withdrawals without deposits (giving back).

Tech is a vital part of our world today and we in Tech should not just be focused on reaping the rewards from it but also ensuring its perpetuation and expansion. That includes helping nurture those in the community who will propagate the system as we go on. That being said, you do not have to wait to be old or on the brink of retirement like the generation before us did, to begin to give back. Start right from where you are and begin to contribute.

Before the question in your mind pops up, I hear you still. *"Am I to go around looking for who to assist or give back to?"* **NO**. Not necessarily. There are various ways you can give back and I share some of them below.

1. **Advice:**

   Be free and willing to offer practical advice to those who seek it and sometimes as well but within careful boundaries. Offer unsolicited advice only to those who may need it. Do not stand aloof or be focused solely on your interests or benefits. If you will rightly recall, it is probable that at some point in time, you may have been stuck or confused and someone was there to offer you advice. Never assume everyone has got it all figured out - many are just winging it. In giving unsolicited advice, it is advisable you approach it indirectly. For example, you could ask the person if they have heard of a certain course or tool or technology and talk about how it could possibly help them.

2. **Assistance:**

Newbies sometimes get stuck on the job. You could assist by clarifying and showing them what to do. There were probably times you were stuck on a task without seeing a way out. You did not want the whole world especially your supervisor or manager - knowing you could not handle things. Remember how badly you needed some assistance back then. If you got it in the past, then freely should you give as you had received. If no one assisted you, recall how crappy that made you feel and how you may have landed a query or some harsh reprimand. Resolve to the best of your ability not to let someone else go through that alone. Help out.

3. **Recommend:**

   Life is about sharing. As you find opportunities that may help those coming behind you, share them. Recommend them for such opportunities too so that they may grow and blossom likewise. In the Tech ecosystem, there is immense power in recommendations. Your recommendation could just be the next big break that person needs, give it.

4. **Share:**

   So, while you may share with others in the workplace, you could extend your influence beyond the confines of the office. A good way to do that is to write or blog. Write about your experiences. If you do not have the desire to run a blog, you could reach out to a wider audience by writing posts that could help them on your preferred social media platform.

"What should I write?" You ask. You may write about your experiences from a technical standpoint e.g. *How to Build xyz*, as well as from a non-technical standpoint e.g. *How I Transitioned to a Tech Career etc*. Other options could be volunteering as a contributing writer to Tech blogs or publications. Find your message and share it with those in the ecosystem. We all need constant information and like they say, *"Knowledge is the new gold"*. So be willing to share some of your gold as a way of giving back.

5. **Speak:**

The beautiful thing about speaking is that it not only enriches your audience, it enriches you as well. Have you, while speaking at a previous engagement, said something that blew your audience's mind as well as yours? As soon as you said it, both you and your audience had a "wow!" moment and for a flashing instant, you wondered where on earth that amazing auditory goodness came from. Talking reinforces the knowledge you have, which becomes part of your second nature overtime. You can give back by speaking at events. Find opportunities to speak at small meetings, conferences, or even with friends and colleagues for a start and then with time, you could get greater platforms to speak on. Big platforms or not, the objective here is to find ways to give back.

6. **Mentor:**

Taking on mentorship is a great and noble responsibility. This is beyond assisting or advising someone as many people think it is. Mentorship is like being a guide and guard to someone. It is committing yourself to the success of another, pouring yourself and your resources (time, energy, wisdom etc.) into that person(s). With mentorship, you help your protégées succeed faster and avoid pitfalls that you may have fallen into in earlier years.

> *With mentorship, success is replicated, growth is accelerated, greater feats are accomplished together in the meteoric fusion of the past and present (ideas, experiences, knowledge, hopes, dreams and resources) that births the future.*

Mentorship is a great way to give back and ensure career sustainability. With mentorship, success is replicated, growth is accelerated, greater feats are accomplished together in the meteoric fusion of the past and present (ideas, experiences, knowledge, hopes, dreams and resources) that births the future.

7. Serve:

There is a saying that you never go beyond your level of service. No matter how high or far you go, never stop serving. Serve those below you and serve those above you. Service is one way to access higher levels you desire to attain. The reward for service is often access to platforms beyond your reach.

> There are levels you cannot access until you serve. We see this play out in almost every sphere of life, even politics - there are always "godfathers" to submit to and serve. It is the same in Tech: you must find ways to serve. Your service sustains your relevance and your relevance always secures your position in any place.

So dear Tech professional, serve your superiors, contemporaries, subordinates as well as the system. Just like we eat and enjoy the fruits of a tree planted in the soil, the soil also needs us to give it back the seeds of the fruits it provides for us so it can reproduce more fruits. Keep this in mind and never stop giving back.

Finally, I hear you again. *"OK Chuka, well said. What next?"* Rinse and repeat

Chuka Ofili

References

1. Carter, S. M. (2020, February 20). The highest-paying jobs in the world. Fox Business. Retrieved April 12, 2021, from https://www.foxbusiness.com/lifestyle/the-10-highest-paying-jobs-in-the-world.

2. Starting a career in Tech after 40? Tips for navigating it all. Skillcrush. (2021, November 16). Retrieved April 14, 2021, from https://skillcrush.com/blog/getting-started-in-tech-after-40/

3. U.S. Bureau of Labor Statistics. (2021, September 8). Computer and Information Technology Occupations : Occupational Outlook Handbook. U.S. Bureau of Labor Statistics. Retrieved April 20, 2021, from https://www.bls.gov/ooh/computer-and-information-technology/home.html

4. Landing a non-tech job in tech: Who's hiring today ... (n.d.). Retrieved April 29, 2021, from https://www.glassdoor.com/research/non-tech-roles-at-tech-companies/

www.ingramcontent.com/pod-product-compliance
Lightning Source LLC
LaVergne TN
LVHW010609160826
845677LV00013B/3332

* 9 7 9 8 3 5 2 2 5 9 5 5 9 *